THE LITTLE BOOK OF QUANTUM SECRETS

DR. PATRICIA VAN PELT-SCOTT, PHD

THE LITTLE BOOK OF QUANTUM SECRETS

How To Get More Of What You Want On Purpose And Without Fail

PATRICIA VAN PELT-SCOTT, PHD

www.drpat.net

THE LITTLE BOOK OF QUANTUM SECRETS: How To Get More Of What You Want On Purpose And Without Fail

DR. PATRICIA VAN PELT-SCOTT, PhD

Published by Wilburn Media, 2024.

---

1. http://www.drpat.net

While every precaution has been taken in the preparation of this book, the publisher assumes no responsibility for errors or omissions, or for damages resulting from the use of the information contained herein.

THE LITTLE BOOK OF QUANTUM SECRETS: HOW TO GET MORE OF WHAT YOU WANT ON PURPOSE AND WITHOUT FAIL

First edition. May 29, 2024.

Copyright © 2024 DR. PATRICIA VAN PELT-SCOTT, PhD.

ISBN: 979-8227322869

Written by DR. PATRICIA VAN PELT-SCOTT, PhD.

Also by DR. PATRICIA VAN PELT-SCOTT, PhD

THE LITTLE BOOK OF QUANTUM SECRETS: How To Get More Of What You Want On Purpose And Without Fail

Table of Contents

DEDICATION

THE SECRET THINGS BELONG unto the LORD our God: but those things which are revealed belong unto us and to our children forever...

Deuteronomy 29:29 (KJV)

THE QUANTUM FIELD can be described as a manifestation of the intricate design of the universe. Just as our beliefs emphasize the presence and activity of the Holy Spirit in the spiritual realm, God uses the quantum field to express His power in the physical realm.

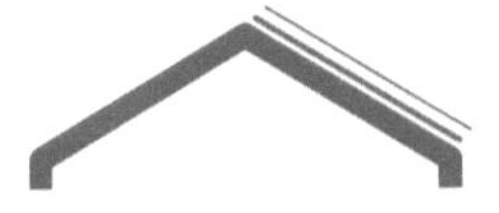

Table of Contents

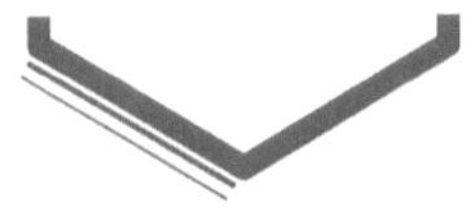

A WORD FROM THE AUTHOR

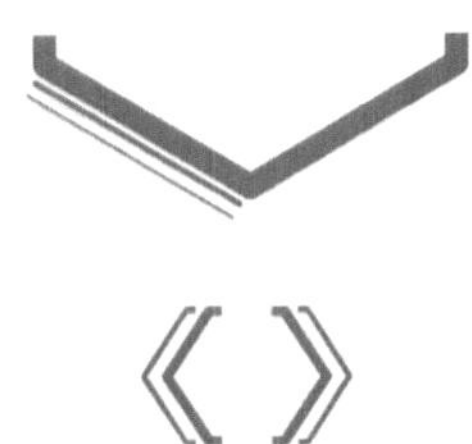

IS IT STRANGE TO ANYONE else besides me that we don't know where we came from? Nor do we know where we go when we leave this world or why we are even here in the first place?

We get a glimpse of the answers to these pressing questions through the scripture. But what happens–if like me–you don't find your way into church until after you are grown?

Society seems to have very little to say about who we are, why we are here, and what we are capable of doing, other than go to school, get a job, get married, have children, grow old, and die. When I first contemplated these essential questions at just 10-years-old, that just didn't make sense. Now, decades later, it still doesn't make sense. I knew there had to be more to the world than what we were being told.

It seems to me that learning the answers to the seminal questions about our origin and purpose should be the primary lessons we learn in preschool, and it should be reinforced all throughout the education pipeline, as it would be the most important thing for us to learn. Yet the establishment provides very little help for us when it comes to answering these essential questions.

So we must learn and share among ourselves, being sure to take advantage of every opportunity for learning that comes our way. Just as all of the major media sources reported on how an orangutan innately knew how to heal a putrefying gash in his face using a medicinal plant, we too must remember who we are.

We must find our way, by first knowing there is more to us than what we've been told; and secondly, be committed to searching for authentic knowledge that is substantiated by many sources, including our inner self. That is exactly what I did to write this book. It's also the same model I used when I established the Advanced Life University and Research Institute. (Learn more @ DrPat.net)

As you take this brief journey through *The Little Book Of Quantum Secrets*, you will see signs and wonders that are showcased through lived experiences, the revelation of ancient scripture, quantum physics, neuroscience, and history. Perhaps, this information will shine the light on who you truly are so you can step into your ideal powerful self and live the influential life you were meant to live.

In addition, I will share a peculiar vision that I had about a trap door hidden in the surface of the earth that I believe will have some relevance for you. Perhaps this information will shine the light on who you truly are so you can step into your ideal powerful self and live the influential life you were meant to live. Let's take this journey!

To learn more about my work, courses, revelations, and conferences, subscribe to my channels on TikTok and YouTube, "A Bit Of Wisdom By Dr. Pat"

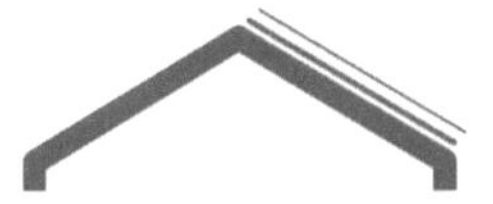

INTRODUCTION:
WHAT IS THE QUANTUM FIELD?

WHETHER VISIBLE MATERIAL or invisible energy, everything we see, feel, or experience–and even things we cannot see are a part of the quantum field.

The quantum field is like the rich nutrient soil of a farmer's field. He plants his seeds in the nourishing soil and with the intense light of the sun and rains of the heavens he reaps a bountiful harvest.

So it is in the quantum field. To produce our dreams and goals, we need intense focus like the sun, and strong emotions like the rain so that the ideas and dreams we plant in the rich soil of our imagination reap a bountiful harvest fulfilling our dreams, hopes, and goals.

The quantum field is like a storehouse where God has laid up everything that we could ever dream of. Just like the Holy Spirit provides spiritual support, guidance, and wisdom for all things godly, so does the quantum field hold all things natural, including our dreams, hopes, and goals, whether they are manifested or yet unmanifested.

The scripture reveals that God has given us all things that pertain to life and godliness. There is nothing we can ever desire that is not already given to us, including a loving spouse, abundance and wealth, strength and wisdom, health and purpose. All of these things are laid up in the quantum field. When we turn our attention toward these things with consistency and strong emotion, they are automatically drawn to us.

"The currency of the earth is money. The currency of heaven is faith. And the currency of the quantum field is focus. Focus on what you want and be blind to everything else and the very thing you focus on will be drawn to you." - Dr. Pat

One of the major tenets of quantum living is Interconnectedness: the inherent connection between all aspects of life, including thoughts, actions, and the external and internal environment, emphasizing the interrelated nature of everything in the universe. So quantum living is not just about science; it is about understanding the deep connections between our spiritual beliefs and the natural, scientific principles that govern the universe.

Just as we trust in the unseen power of God to guide and shape our lives, quantum physics reveals the hidden mysteries of how the universe operates beyond what we can see with our eyes.

In our everyday lives, we experience the power of faith and intention – believing in miracles, speaking words of positivity, and trusting in divine guidance. Quantum living expands on these concepts by showing us that our thoughts and intentions have a tangible impact on the world around us as well.

Just as one might preach about the power of prayer to manifest blessings and miracles, we also understand that those blessings and miracles are laid up in God's storehouse. *The Little Book of Quantum Secrets* takes us further. It is here that we learn how to access those blessings through quantum living.

Quantum living teaches us that our consciousness continuously influences the very fabric of reality. It is so important for us to understand this. Our whole world revolves around our consciousness. Don't leave this book without grasping this essential point. Your internal and external environment is fueling your conscious reality. Look around at the environment you're in right now. Is it suitable for a person that desires to achieve what you want out of life?

If not, go to my website *(DrPat.net)* and download the important guide that I created, entitled, *"How To Raise Your Game In The Next 24 Hours."*

It is a fact that we live in a quantum world. With that knowledge, we must ensure that we are building a life that flourishes in a quantum environment. That's a decision that you can make today!

It is a confirmed fact that in order to get the full benefit out of an experience, we must first be conscious. It is through consciousness that we tap into the limitless potential within ourselves and align with the universal energy to co-create the outcomes that we desire, including true love and marriage, abundance and wealth, health and strength, and hope and purpose. You're not going to just stumble up on these. It takes conscious effort.

This *Little Book Of Quantum Secrets* invites you to find your own personal power and capabilities by exploring the intersection of faith and science, embracing the wonders of the unseen realm while grounding your beliefs in the transformative power of spiritual principles. It's about empowering yourself and all those that you love with intention, purpose, and divine connection in every aspect of life, including finding true love, building substantial wealth, and fulfilling your lifelong purpose. So Let's Go

CHAPTER 1 -

THE QUANTUM FIELD IS HERE TO STAY: EMBRACE THIS TRUTH

LET'S START HERE:

"... as for knowledge, it will pass away [it will lose its value and be superseded by truth]." 1 Corinthians 13:8 (AMPC)

That simply means that our story is still unveiling; we do not have the whole truth yet. So as you journey through life, you have to be open to letting go of beliefs you once held as true when you discover they aren't.

THE QUANTUM FIELD IS HERE TO STAY

The quantum field has always been and always will be. I know it's getting a lot of attention now. So it may seem like a temporary fad; but trust me, it is not. It's just that knowledge is now being superseded by a major truth that we have been blind to during most of our lives.

Our collective consciousness as a society has expanded; so now we see more of the truth. With this expansion, we gained a deeper understanding of how the universe operates and how to use its tools and laws to get more of what we desire in our lives.

FOR YEARS, I WANTED to be a scientist. I wanted to understand the world. I wanted to know what we are and how we were made. I ended up being called into ministry and nonprofit management; but I never lost my hunger for the authentic depths of truth and understanding of the world around me through science. I could see there was a marriage just waiting to happen between my preaching the gospel and teaching of physics and biological science. And voilà just like that, here I am writing *The Little Book Of Quantum Secrets*,

while coaching and training others how to gain mastery over life, love, and conditions. Life is Sweet! Now let us continue in the quantum field...

Before scientists had a clear understanding of the quantum field, they used different ideas to explain how things work in the physical universe. One idea they had was called the "ether." People thought this ether was a kind of invisible substance that filled up all of space and that it helped light and electromagnetic waves move around.

But later on, Albert Einstein showed that the ether—as they described it—likely did not exist because the experiments using the idea of ether didn't line up with what was already known about how larger scale objects, like how planets and galaxies interact in space (*aka Einstein's Theory of General Relativity*). So, scientists stopped using the idea of the ether and sought new ways to describe how things work in the universe.

By the early 1900s, scientists finally figured out how tiny subatomic particles and forces work. They call this line of studies, quantum physics aka quantum mechanics. And they use the term, "quantum fields" to help us understand how these subatomic particles interact with each other and follow the rules of both quantum mechanics (very tiny particles) and *Einstein's Theory of General Relativity* (large scale galaxies). With that congruency, a big door swung wide open, paving the way for a flood of new ideas, technologies, and inventions.

So as it turns out, the quantum fields are like a detailed blueprint that shows how everything fits together in the universe, both large and small. So, instead of thinking about the ether, scientists now focus on these quantum fields to help us understand the building blocks of everything around us, including our consciousness, capabilities, and our personal power. You may accept this as a reality because of the expansion in your own consciousness. But don't be dismayed if others outright reject this idea completely without any further research or study.

Just note what Dr. David Hawkins, MD, PhD said, "We have frequently pointed out that man is unable to observe or recognize an event until there is a prior context and language for naming the event." Dr. Hawkins' words ring true in every area of life. It is almost impossible for us to understand anything that we do not have context for. It's not that we can't understand it. It's just that we have stopped allowing our mind to wander.

We've forgotten 1 Corinthians 13:8 (AMPC) "... *as for knowledge, it will pass away [it will lose its value and be superseded by truth].*" So we limit our imagination to include only what we have come to believe is possible; everything else seems out of context. Thus, we struggle with anything new.

BUT CLEARLY YOU ARE not the norm because right now, you have an advanced copy of *"The Little Book Of Quantum Secrets"* in your hands. And that says a lot about who you are! Now let's continue to expand our understanding about all things quantic!

Here are a few quantum principles that you should know. I've added supporting scripture for each of them:

1. Wave-Particle Duality: Particles can exhibit both wave and particle properties. This is akin to Jesus' walking right through the angry mob that aimed to throw Him over a cliff one day after He preached a sermon at church that they didn't like. *See Luke 4:28-30.*

2. Superposition: Particles can exist in multiple states simultaneously. This is akin to the omnipresence of God as well as how we, operating in the gifts of the Spirit, (i.e. the word of knowledge, prophecy, out-of-body experiences) can also be in more than one place at the same time. See Jeremiah 23:24; II Kings 5:21-26

3. Uncertainty Principle: Formulated by Theoretical Physicist, Werner Karl Heisenberg, it states that the more precisely one property is known, the less precisely the other can be determined. This is revealed through scripture in that many things can only manifest in the physical world when we believe it's possible. See Matthews 9:27-29; Mark 6:5

4. Entanglement: Particles can be entangled, and the state of one particle instantaneously influences the state of the other, regardless of distance. Scripturally, this is clearly seen through the process of giving and receiving as well as sowing and reaping. When we give or sow, we immediately are entangled with the return of abundance of whatever we sow. See Galatians 6:7.

5. Quantum Tunneling: Particles can pass through energy barriers that classical physics predicts they shouldn't be able to. We see this throughout the scriptures as Jesus walks on water, through walls, and even enters death and resurrects himself. He

follows on to tell us that—though untapped for now—we have the same power. John 14:12

Wow! There is so much more to us than meets the eye! My 10-year-old self is shouting for joy right now because just like you—I knew there was more to life than what we were being told, and most assuredly, more there is! As providence would have it, my detour from science into the ministry, academia, and nonprofit management granted me the foundational understanding that I needed to handily teach the science of the quantum world. And on top of that, it dovetails well with the spiritual world.

> ● Looking back, it is clear that the trajectory of my life couldn't have been written better. I am here teaching science, spirituality, and quantum living—a powerful combination that can help any seeker to steer their lives to a higher realm of success on purpose and without fail. Now let's go deeper.

As we apply the quantum principles to our life, we learn that we possess tremendous capacity that gives us the power to gain mastery over our thoughts, emotions, and feelings. In addition, we get mastery over matter. This aligns with the Word of God, *"Be fruitful, multiply, take dominion over earth and subdue it..."* Genesis 1:28 *(paraphrased)*. To take dominion or to subdue anything requires mastery.

This is exactly why I hold conferences and teach courses at the Advanced Life

University. We must come into mastery over life, or we will continue to live subserviently. And I'm sure that's not your goal in life. Right?

Adhering to quantum living principles brings us into a more conscious, interconnected, and intentional mindset that not only frees us from perceived limitations, but also empowers us to deliver others! We get access to everything we buy, we come to understand our birthright and how to use the power we have been given.

As we come into greater consciousness about the power we possess, we can command the resources of the earth and they will obey us. Yes!

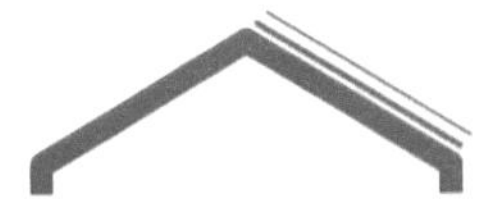

CHAPTER 2-
UNLOCKING QUANTUM POWER

I'VE ALWAYS LOVED RESEARCH and learning. And that's how I came into the knowledge of what scientists are now calling the quantum field. Even though I didn't have the name "quantum field" in my vocabulary, I've known what makes it work for nearly a decade. In fact, it's through my understanding of the quantum field that has led to my high achievements and massive success, including personally earning over a million dollars and owning 2 multi-million-dollar businesses. More on that later.

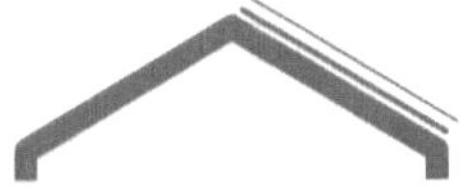

Here's a quick true story.

I STILL REMEMBER THE day when I was sitting in an audience of approximately 10,000 people listening as several multi-millionaires came to the stage one-by-one and shared their story about how they achieved their wealth-status.

After listening for a while, I decided there was nothing that they had that I didn't have—that is, except the money. So I set a goal to reach a pinnacle of success in my business in record time because I wanted to get on stage to speak and I knew I deserved it. Plus getting on that stage meant I would win a brand-new drop-top Bentley, and I wanted that too! I was working hard and doing what I needed to do in order to reach that level of success. But soon I realized that my progress was too slow to reach my goal on the timeline that I had set for myself. So I started doing what I do best; I started doing qualitative research.

I set meetings with 7 people that had already reached the goal that I was striving to achieve. Each one went through the motions of describing the necessary steps to reach the goal. They were helpful. But those were things I was already doing. But one, a Mr. Tupac Derenoncourt said something that none of the rest said. And with his words, my life took an explosive upward ascendance.

He said, *"Pat, you know it's all mindset, right?"* I nodded. But truly that was a revelation for me. I had maintained a positive mindset for spiritual things for decades. Even calling those things that are not, as though they were. But I never ventured into the business world with the power and command of mindset as I did in the spiritual world. For some reason I approached wealth, love, and success with passiveness. I had a "don't pursue" and "wait until God gives it to you" mindset for those things.

But after that conversation and from then on, I began to intentionally engage with the quantum field not just for spiritual success, but for financial wealth and for soulmate attraction too.

Within months of that encounter, I went from earning just enough to get by to eventually generating over One *Million* Dollars personally and becoming a founder of what would become two multi-million-dollar businesses. I went from not having a car of my own to owning that brand-new drop-top Bentley I wanted and a BMW truck to boot in the same year. Then I bought my first home for cash, and eventually acquired 10 additional properties.

Next I had to face a strong opponent for the coveted Illinois Senate Race, and I beat him handily against all odds. Then I met the man of my dreams, and we were married four months later and are celebrating eight years of blissful marriage this year.

All of that was triggered by the life-changing moment that I came to understand the power that I wielded in the spiritual world, also works in the natural world.

> *"Sometimes, you are just missing a simple piece of information. And when you get that one piece, everything changes."* -**Dr. Pat**

Now I want you to know how to engage in the field to get what you want because, like me, you might only need this single revelation to cause the abundance of the seas to be converted to you. I've added an exercise at the end of this chapter to help you navigate the quantum field. Don't miss it. If you need an accountability partner or more help, contact me at ***DrPat.net.***

Now here's a scenario to remember:

"The Quantum field is like an invisible ocean that fills the entire universe. And just like the waves in the ocean are caused by underwater disturbances, so do we cause waves with thought, beliefs, and words." -Dr. Pat

Yes, we stir the infinite field of possibilities with our mindset, which is fueled by our thoughts, beliefs, and customs. But it takes additional steps to get the fruit of those thoughts to manifest in physical form.

Powerful Keys

I KNOW WE PRAY. I KNOW we fast. I know we work for God. But understanding the value of meditation, breathwork, and visualization are crucial, as these are powerful keys that successful people never ignore. And I do mean never! For some of us, these keys might contradict what we've been taught as truth. But if we just look, we will see that all of these are aligned with the scriptures.

We know that visualization is the same as *"write the vision and make it plain"* Habakkuk 2:2-3

Then the LORD answered me:

"Write down this vision and clearly inscribe it on tablets, so that a herald may run with it. For the vision awaits an appointed time; it testifies of the end and does not lie. Though it lingers, wait for it, since it will surely come and will not delay...."

IN OTHER WORDS, IF we visualize the things that have already been promised to us, they will manifest. Just don't let it go. Hold the vision firmly in your mind. We know breathwork is the same as God breathes into man and he became a living soul. We know that meditation is being still or quiet when we are waiting in the presence of God.

So let's not stumble over words today. Let's allow ourselves to continue this journey without fear. Let's go!

Let's get an understanding of this quote:

"The currency of the earth is money. The currency of heaven is faith. The currency of the quantum field is focus. Focus on what you want and be blind to everything else and the very thing you focus on will be drawn to you."

Picture this:

A DRY, DESOLATE LAND, where the scorching sun beats down mercilessly, and the earth cracks under the weight of relentless drought. This was the reality for the Navajo people, spanning across the arid landscapes of New Mexico, Utah, Colorado, and Arizona. Amidst this bleak landscape, a crisis unfolded. Crops withered, animals perished, and the Navajo found themselves locked in a desperate struggle for survival. It was as if the very essence of life was slipping through their fingers, leaving them grasping at shadows of hope.

But in the face of this existential threat, the Navajo leaders refused to surrender to despair. Instead, they turned to an ancient prayer—a sacred invocation passed down through generations, a beacon of hope in times of darkness. This prayer, known as The

Beauty Prayer, held within it the key to unlocking the path to abundance.

The Beauty Prayer was a call to focus on the beauty that surrounded them, even in the midst of turmoil. It urged them to seek out moments of beauty, in every situation they encountered no matter how fleeting or hidden. It was a reminder that beauty could be found even in the harshest of landscapes, if only one dared to look.

And so, with hearts heavy yet spirits undaunted, the Navajo people embarked on a journey of discovery—a journey guided by the words of their ancestors, leading them towards a glimmer of light amidst the darkness. And lo and behold, as they embraced the wisdom of The Beauty Prayer, they found solace, strength, and resilience in the face of adversity.

The Beauty Prayer became their lifeline, their anchor amidst the storm. It infused their days with purpose and their hearts with hope, empowering them to not only survive but thrive in the face of extinction. Once again, defying the odds and writing a new chapter in the annals of their history the Navajo people emerged victorious.

So let us take heed to this timeless tale—a testament to the profound power of words, the resilience of the human spirit, and the enduring beauty that lies within us and all around us just waiting to be discovered, even in the darkest of times. I hope you enjoyed this. I have added The Navajo Beauty Prayer for your enjoyment! You will find it in **Appendix B**.

Now as promised, here's the powerful exercise to consciously engage the quantum field.

Follow the steps below to see how the quantum field is at work in your life.

Focus on at least one major life outcome that you have not achieved yet but are tired of living without it. Try to be as realistic as possible. But pick something that has been slightly out of your reach. Now let's move that goal into the space where dreams manifest into the physical realm!

Write the goal down with a 12-week timeline and place it in a prominent place where you can see it every day.

Next:

MANIFESTATION ARENA: List up to 3 goals/dreams that you need to manifest over the next three months that you feel are necessary to help you achieve that life outcome that you are tired of living without. Add a brief statement about why you no longer will continue to tolerate life as you have in the past.

MOVEMENT MOMENTUM: List three to five activities or things you will have to move in order to achieve each of the three goals that you listed above. Add a timeline that reflects your 12-week deadline.

MEDITATION MASTERY: The Key to Unlocking Quantum Power is focus. Most people would think that you should meditate on the life outcome you want to achieve. Instead, meditate on the process and visualize yourself following the process you outlined above.

Pay attention to what you see, feel, and think as you journey in the unseen world of your life outcome. Make corrections if you see anything that needs to change or pivot to make sure you achieve your goals in the time allotted. Remember the words of

Tupac Derenoncourt. It's all mindset. So what you think about the most is what happens. The currency of the quantum field is focus. Be blind to everything else and the very thing you focus on will come into the light.

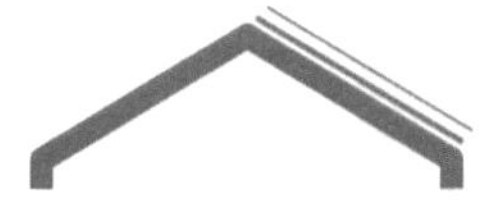

CHAPTER 3 -

THE MIS-EDUCATION OF THE MASSES: SECRETS OF THE QUANTUM FIELD

SOCIETY, AS A WHOLE, has perpetuated the idea that there is lack and scarcity throughout the world. This notion goes all of the way back to Charles Darwin's seminal work published in 1859, *"On The Origin of Species"*, where he espoused his Theory of Evolution, which included the concept of **"survival of the fittest"** and more.

Though it was not Charles Darwin's intention, his theory was used to promote the idea that some races were superior to others, and that only through competition would the fittest among us flourish. Under this theory, we would need to compete with each other for survival. And that only the fittest would survive.

This idea of competition, strength, and force would be taught at our most esteemed universities, embraced as truth by the government, pontificated by media sources for over 100 years, and eventually embraced as common knowledge across all of society. Though these ideas became known as truth, in reality there is no lack of resources anywhere and there never has been. Thank God that knowledge is now being superseded by truth.

The whole universe is quantic, which means that everything in the universe shares the same building blocks. Every person and everything, whether visible or invisible is made from the same substance--atoms, and there is no lack of atoms in the universe. Quantum physics is the study of atoms that have been broken down to subatomic particles, such as neutrons, protons, and electrons. These are the building blocks of all things. Everything, including us, is made of subatomic elements.

And of course, with these new revelations about our universe and about our existence more is being revealed every day. We are surrounded with infinite potential waiting to be brought into the physical world; it's called The Quantum Field. That's why all things are possible naturally and spiritually.

Furthermore, according to the Word of God, everything already exists.

"What has been will be again, what has been done will be done again; there is nothing new under the sun." Ecclesiastes 1:9 (NIV)

THE NEW LIVING TRANSLATION Version captures it best. It states:

"History merely repeats itself. It has all been done before. Nothing under the sun is truly new. Ecclesiastes 1:9 (NLT)

Since everything already exists, that means that the reality we're experiencing right now is simply a reflection of what we believe, think, and say is possible. Clearly, the wealthy and their offspring have known this for centuries. And they've used it to procure more wealth and resources for them and their families all while our education system has kept the masses cloaked in the dark ages of the past.

For an entrepreneur, ancient scripture, quantum physics, and the law of attraction can be powerful tools for achieving business success. Essentially, your thoughts, beliefs, and intentions shape your reality, influencing the opportunities and outcomes that come your way. In practical terms, this means that focusing on maintaining a mindset of abundance, you will attract opportunities, resources, and people that align with your vision for your business.

By visualizing your desired outcomes, setting clear intentions, and taking inspired action towards your goals, you can create a positive feedback loop where your success breeds more success so that you can achieve profound results in every area of your life. But what happens if rather than seeking success, you drift through life not intentionally seeking anything good or bad? Nor taking any action aligned with your dream?

"Therefore, I say to you, all things for which you pray and ask, believe that you have received them, and they will be granted to you." Mark 11:24 (NASB)

WHAT COULD BE PLAINER than that? The energy you put out into the universe, whether positive or negative, will multiply similar energy back to you. Period. If you drift, you get whatever is leftover by those that already decided what they want to achieve. Drifters never win because they are aimless.

CONSIDER THIS:

"From the days of John the Baptist until now, the kingdom of heaven suffers violence, and the violent take it by force." Matthews 11:12 (KJV)

This tells us that the kingdom of heaven is not something that you can drift into. Nor can you just drift into the life you deserve.

Once you learn how to get violent about the life you truly deserve, the needs of your life will be chased away just as the morning sun

CHASES AWAY THE DARKNESS of the night. And it will be replaced with the power and resources that you need right now! Are you ready for this?

HERE'S ANOTHER TRUE STORY

IT WAS FRIDAY THE 13TH, a day that Steve will never forget. He and his good buddy Tiny had been moving some heavy equipment for the Hawaiian Electric company in Oahu, Hawaii. They had done this type of work together for a number of years, even during the period that they had served as soldiers in Vietnam. Everything was going great and there was nothing unusual about this day. In fact they had made plans to just chop it up and have some leisurely time at the local bar a few hours later with friends.

Steve was flying the chopper as usual, and Tiny was laying the steel beams in place as Steve lowered them to the ground. According to Steve, the day was about done. He only had six pieces of equipment to transport to the area. And now with only a few more pieces to move, Steve thought the day was all but over. But that's when all hell broke loose.

Steve recounts, "I had moved two of them and was coming back for the third one when the helicopter tried to go into a full left-hand turn without warning." Steve had never seen the helicopter do that before. Once he got it under control again, Steve made the decision to come back and land the helicopter away from where Tiny and the other workers were. But instead of landing, the helicopter went out of control again.

Steve recounts, "I was too close to the ground, didn't have a chance to do anything. All I can remember is the helicopter pitching to the left and up, and then that was all."

With a violent crushing sound, the chopper crashed into a stream. Steve was face down in the water. His head pinned underneath thousands of pounds of twisted metal. It was a devastating scene. With no way to free Steve, bystanders shouted, "Call an ambulance." Get help!!

But there was no time to wait for rescue equipment to arrive. Steve was drowning and for a minute it seemed as if there was nothing Tiny or anyone else could do about it. With no help in sight, Tiny's friend lay drowning right in front of him.

Tiny scanned the wreckage searching for something... anybody... anything that he could use to help his good friend, as Steve lay lifeless in the water. For most people, this would have been a time of great agony, grief, and pain. Except Tiny didn't take the path of most people. Nor did he look to his past for help. Tiny went within. He went deep within and that's when he got a glimmer of hope.

Tiny recounts, "I don't know what made me do it, but I just grabbed the chopper to see if I could move it. And the first time I tried it, it budged a little. Then I guess the grace of God picked up the next section. The next time I went down to pick it up, it moved.

I had it lifted high enough for one of the bystanders to get Steve's legs out and then his whole body."

Then Tiny began giving Steve mouth-to-mouth resuscitation while repeatedly begging him to not give up and to stay alive and encouraging his friend to keep fighting until medics could airlift him to the hospital. Tiny told his friend, "You're all right. They're coming now. The ambulance is coming. You will be fine."

This story may seem unbelievable. But it is true and there are countless YouTube videos online that captured and documented this harrowing episode of Steve and Tiny's life.

Steve survived. He had a broken ankle, a severe concussion, and his arm was nearly severed. But because Tiny carefully packed Steve's mangled arm in ice, the doctors were able to save Steve's arm. Steve credits Tiny, he said, "Without a doubt Tiny saved my life. When I finally was able to watch the videotape after the crash. I saw what he did. He jumped into the water without hesitation, reached in and tried to get me out, not even considering whether the thing was going to explode into a million pieces. Then he grabs the helicopter and actually physically lifts it, which is humanly impossible to do. And yet he did it. And that's a heroic act."

To that comment Tiny replied, "I don't think I'm a hero. I'm not being modest. But when you go to help your friend, you don't run there to get recognition. You run there to help your friend." Tiny refused to stand there watching his friend Steve drown to death while being crushed under the helicopter. So he took decisive action.

WHAT IS HAPPENING IN your life right now that you cannot stand to watch anymore? Make a list of those things now.

"When we shift our perception, our experience changes." – Lindsay Wagner

Just as Tiny commanded the resources of strength, wisdom, and power, so will you as you learn and meditate on the innate abilities you have been given in the seen and unseen world.

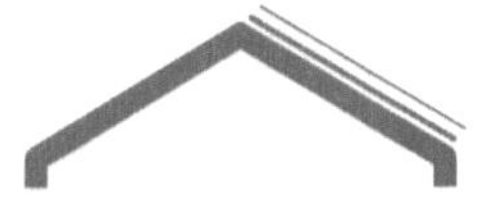

CHAPTER 4 -

EXPAND YOUR VISION: THINK BIG, ACHIEVE BIGGER

THE LADY ASKED, *"WHY is it that every time I try something, I never get big results?" I told her you never get big results because you only attempt to do small things."*

THE QUANTUM FIELD: GOD'S STOREHOUSE OF PHYSICAL RICHES HAVE BEEN GIVEN TO US

Your words and thoughts have power over matter whether you recognize it or not. If the word of man did not have power over matter, God would never have commanded Adam to take dominion over the earth and to subdue it. Without power over matter, how could Adam possibly take dominion, have authority, and subdue the earth along with all of its resources?

We were all born with this level of power. And it has nothing to do with being good or worthy. We were born with this power. To help you understand this, consider Cain's story. He killed his brother, dissed God, and then got himself banished out of the presence of God. It would seem that would have been the last we heard of Cain. But it was not. Cain's life showcases the innate power over matter that we all have. His story is captured in the book of Genesis.

> **Genesis 4:16-17 "And Cain went out from the presence of the Lord, and dwelt in the land of Nod, on the east of Eden. 17 And Cain knew his wife; and she conceived, and bare Enoch: and he built a city, and called the name of the city, after the name of his son, Enoch."**

CAIN DID NOT GO OUT and apply for Section 8 low-income housing; he did not seek the least paying job due to his record. Instead Cain went out and built a city and named it after his son. Even after he murdered his brother, Cain had no problem producing children that had gifts and abilities to create and to build. Why? Why was Cain able to still prosper in the way he did in spite of the fact that all the things he did were so contrary to the Word of God?

IT'S BECAUSE THE GIFTS and calling of God come without repentance and the quantum field (aka God's storehouse) is available to us all.

Everybody has these gifts, every one of us has this calling. But some of us pay more attention to poverty and what we feel we can't do than to prosperity and the things we can do. Cain's ground was cursed; but his hands weren't.

Some of us groom our talents until they become skills. Others mainly focus on and pine over the things and abilities that are out of reach. We must understand that Cain did not allow his history, his background, or anything else stop him from being able to do great things.

We have to ask ourselves why it is that even the demons got their prayers answered when they made their request.

"Please Mr. Jesus, don't make us go out of the city to another country, let us go into the pigs. And Jesus honored their request." Matthews 8:31-32 (NASB)

When Ahab the wicked king cried out to God and said *'Lord I don't wanna die like this. I want a chance.'* God sent the prophet right back to him and told him *'I'm going to give you that chance.'* 1 Kings 21:27-29 (NKJV)

Why were they all able to do these things? Why was Cain able to do what he did? Why did the demons get their prayers answered? Why?

The reason why is, they took the time to ask. He said ask and you shall receive, seek and you shall find, and knock and the door shall be opened. He's simply saying the reason why you have not is because you ask not. And when you do ask you ask amiss. So it's not that you can't have it. It's just that when you ask for what you want, know that all of these things belong to you. That way you won't come with doubts and fears when you ask for what you want.

The quantum field is like an invisible ocean that fills the entire universe. Waves and ripples are a common sight in the ocean. They are formed by disturbances projecting underneath the water. And so it is with the quantum field.

Just like the disturbances under the water, our thoughts, emotions, and beliefs all project into the universe and create waves of potential in the quantum field for better or for worse.

Whether visible material or invisible energy, everything we see, feel, or experience—and even things we cannot see are a part of the quantum field. The true basis for our reality is magnificently powerful. This *Little Book of Quantum Secrets* is here to illuminate the path to your storehouse of infinite potential.

We cannot come to the universe confused about what we want. We have to get right down to what we want and why we want it. And it has to be important enough for us to put our attention on it and keep our attention on it. We have to stop wondering, does

God want me to have this or does God want me to have that? All of those should be bygone conclusions. Yes God wants you to have this and that.

HERE'S A PERSONAL STORY:

During the month of March, I set a goal for two things: 1) seven days away in Florida at the Pritikin Longevity Center so that I could spend some time thinking, praying, and resting; and 2) I wanted our deck repaired and painted.

Well it turned out that repairing and painting would not be enough; the deck needed to be completely rebuilt. The cost shot up substantially from $3,500 to more than $10,000. And the cost for The Pritikin Longevity Center priced out to be over $8,000, which was almost double the amount that I had paid in the past. So that was far more than I was prepared to pay for a one week stay there.

I mentioned my goals to my pastor, husband, and a few others. My pastor let me know that he had some guys working for him on another construction project. But he would bring them over to repair and paint my deck at no charge. I tell you, I celebrated that. But when they got there they discovered that the entire deck would need to be rebuilt.

AND TRUE TO HIS WORD, my pastor had the work done at no charge to me. Yes!

As the month of March rolled on and I had not been willing to spend the $8,000 for one week at Pritikin, I told my husband Gene, if I can't go to Pritikin this time, I am gonna bring Pritikin to me! I would just follow the process that I learned the last time I stayed there.

Gene said, *"Don't be so quick to give up on your dream. You still have eight days left in the month. Let the month play out. You don't know what might happen."* He was right.

On March 30th a friend called who didn't know anything about my goals for the month of March. Yet, she offered me a free seven-day cruise on Royal Caribbean's "Icon of The Seas", the newest ship and the biggest ship on the ocean. As it turned out, it launched from Florida, the very state that I had planned to fly to for my stay at Pritikin! You would think that it was a slam-dunk decision since it was everything that I set out to get done in the month of March; but it wasn't.

The reason it wasn't a 'slam-dunk-YES' was because of me. I was in the way. It was me and my custom of inquiring about whether God wanted me to have a thing or not. And I hadn't gotten an answer that day, so everything was held up. Since I didn't respond with an immediate yes, my friend let me know that time was of the essence. We were running hard against a deadline for her to register me on the cruise.

While I was thinking and praying about whether I should go or not, these words came to me, *"If it's not a no, then it's a yes. I've already given you all things that pertain to life and godliness. I want you to have all of these things and even more. So don't ask me whether you should accept the very thing that you already prayed for."*

In other words, when the things you pray for manifest, don't blink!

Wow! So of course, the next morning I graciously accepted the trip as it was another manifestation from God's storehouse—the quantum field.

As I requested, I got the cruise and the deck rebuilt in the month of March and I got them both at no charge!

Over time, I have advanced my thinking and understanding of how to use the power that belongs to us. All of these things have always belonged to me and you. I asked. And just like that my request was granted.

What are your dreams? What are the goals you hold dear?

Write to me here: *Hello@DrPat.net*

Tell me about your dream and how you're using the teaching in this book to help you achieve them (See Exercise One in Chapter two).

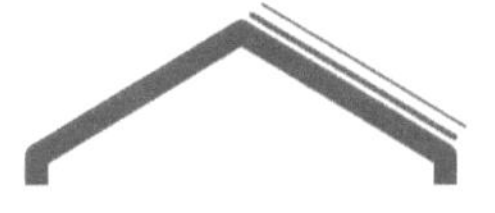

CHAPTER 5 -
MANIFEST YOUR DESIRES: YOUR QUANTUM ORDER EQUALS YOUR REALITY

THE QUANTUM FIELD IS LISTENING

In some states, it is illegal to record someone without their knowledge. But men's laws do not govern the quantum field. The quantum Field is always listening to you. Everything you say, think or do impacts your world and everything around you.

We see this displayed in the Quantum Entanglement Principle: Particles can be entangled, and the state of one instantaneously influences the state of the other, regardless of distance.

Also in the scripture: what you sow, you will also reap, meaning your actions are always entangled with a response.

"Do not be deceived, God is not mocked; for whatever a man sows, that he will also reap..." Galatians 6:7 (KJV)

When you understand that your words and thoughts have power over matter, you will change the things you say and think.

Surprisingly, you can even change all forms of nature with your words and thoughts too. Your thoughts don't just stay in your head.

No one has done more extensive research on this subject than The HeartMath Institute. HeartMath has spent decades studying how the heart intentions impact our lived experiences. The HeartMath image below reveals that each of us have a measurable magnetic field generated from the heart that can be detected up to 3 feet away from the body. And that magnetic field beams our thoughts, beliefs, and emotions into the quantum field. We are communicating with the universe through the energetic frequencies emitting from our heart, all day and every day.

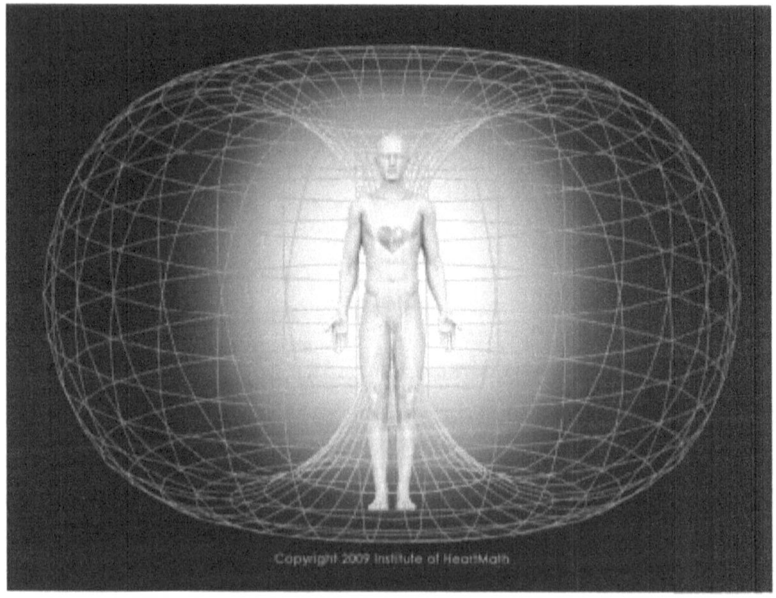

FIGURE 1: THE HEART'S magnetic field, which is the strongest rhythmic field produced by the human body, not only envelops every cell of the body, but also extends out in all directions into the space around us. The heart's magnetic field can be measured several feet away from the body by sensitive magnetometers. ("Chapter 06:

ALL OF YOUR REALITY is seeded by what you put out into the quantum field. To quote Henry Ford, *"Whether you think you can, or you think you can't – you're right."*

This quote helps us understand how our attitude plays a major role in determining our success or failure. Why? Because the quantum will see to it. It's not here to punish or reward. It is simply here to follow through with what you say you want. Your words are so powerful in the quantum field. They can change the trajectory of your life as well as others.

> *"In the same way, the tongue is a small thing that makes grand speeches. But a tiny spark can set a great forest on fire."* James 3:5 (NLT)

Consider this experiment:

Ikea, the furniture store, proved the power of our words over nature. They put two plants in a school. Both received the same light and water. But the students were instructed to bully one of the plants several times a day by hurling insults, such as, "You Look rotten!" "Are you even alive?" You smell terrible!" and such like. While they were instructed to give praise to the other by offering compliments, such as, "You're beautiful!"; "You're flourishing!", "I love you!", and several similar statements. After 30 days, the bullied plant was all but dead. While the celebrated plant was flourishing.

Just like you can bully a plant and kill it, you can kill the things that are around you with your words, including you, your hopes, and your dreams. Every word, thought, and belief engages with the quantum field. You have power! With that knowledge, it is high time to take the reins of your life, and steer your words, thoughts, and actions toward fulfilling your life goals and dreams.

REPEAT AFTER ME, "Down with vagueness, Up with clarity!"

Remember, your words have power. But vagueness does not work to your advantage. If your description of an outcome is vague, it's highly unlikely that you will get what you want. So "down with vagueness, up with clarity!"

Here's An Allegory Worth Remembering.

A woman went to a restaurant one day and sat down to order something to eat. She was hungry and thirsty. But when the waiter came over and asked her what she wanted to drink, she said, "whatever." He went on to ask, would you like coffee? Again, she said "whatever." Because the waiter became more persistent, she said, "just bring me anything to drink."

He brought her some tea. She said, "I don't like tea." And though she didn't like it, she drank it anyway, complaining all along the way.

The waiter returned and asked, "What would you like to order for dinner?" She started off saying I don't like chicken. He said, "Yes, so you don't want chicken." So what do you want? She said I don't like a lot of things." The waiter began to point out various items on the menu. He said, now what would you like to order? She said, "I don't know yet." When he came back for the third time, and asked, "What would you like to order?"

She said, "Just give me whatever you have as a special today." So he brought her fried chicken. She said, "I don't like fried chicken!" But again she ate it anyway, complaining all along the way.

How many of us are continuously complaining as we drift through life? Are we still settling for whatever comes to us instead of getting really clear about what we want to happen in our life and refusing to settle for those things that do not match our true desires?

If that's you, today can be the day that you change the course of your life by simply acknowledging and activating the innate power you possess to create the reality you want to live in.

When you get ready, you can upgrade your life. I mean you can seriously upgrade everything concerning your world by first getting crystal clear about what it is that you really want, envisioning the value and benefits of that new reality, and taking steps aligned with the new you.

When we are not clear about what we want, we tend to get whatever is floating by at that time. The quantum field is conscious, and it responds to clear and concise instructions..

The double minded man is unstable in all of his ways. So you must get clear, concise, and consistent about what you want and stop accepting what you don't want. As my coach told me many years ago, "Patricia, you get what you tolerate. If you're tired of getting that, stop tolerating it."

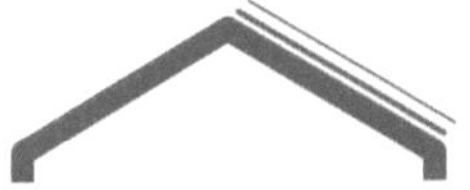

Remember, the quantum field is listening.

Exercise:

MAKE A FRESH COMMITMENT to that one life outcome that you are tired of living without. Continue to follow your plan and create a gratitude journal on each of the following 30 days, being fully assured that what you can't live without anymore will manifest.

BE SURE TO WRITE TO me about your success, at *Hello@DrPat.net*.

I am excited about your future. I will read and reply to as many of your stories as I can.

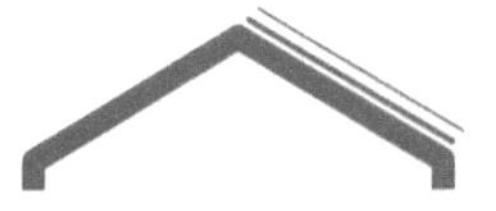

CHAPTER 6 -

THE REVELATION OF THE TRAP DOOR: IT'S TIME TO BE SEEN

ACCORDING TO THE SCRIPTURE, many are called, and few are chosen. I believe if you're reading this book you must be one of the chosen. You are chosen to do something specific. You have significant power and insight that other people don't have. You've always been different from other people. That's why—as promised—I must tell you about the vision that I had regarding the trap door.

There was a trap door on the surface of the earth. The trap door led to a tunnel where thousands of people had hidden themselves.

I opened the trap door and led the people out from under the earth where they had hidden themselves. They all had a preponderance of gifts and talents. But they had hidden them because they were so different from most of the population that they were afraid of the persecution they would suffer because of those differences.

They were afraid that they would be ostracized and ridiculed by the general population. But I opened the trap door because now it's time to come forward. And it's safe now. It's time to step into the space that God has ordained for you to be in.

The world is headed somewhere; and it is not good. The world is barreling toward a certain future. It will take the true sons of God to step up to cause the world to detour before it reaches the end of this cycle.

Remember, all through history we see that the right people stood up at the right time and in the right place and changed the trajectory of the world. It's time for you and me to stand up. I'm sure you may be saying, "I don't have the skills or the ability to talk, write, speak, or even to be around people."

Many people that are part of this group don't even like being around people at all because it's really uncomfortable to be doing a performance all day. Lol!

But it's necessary now. So what do we have to do so that we don't have to perform all day and every day? We have to come to mastery.

That means learning how to master your emotions in the midst of the circumstances and conditions you're in right now.

KNOWLEDGE OF THE QUANTUM field should help you, especially since you've known about it (maybe by another name) since childhood.

You are on a mission, and you are not alone. So you must come out of hiding so that you can achieve the dreams that's in your heart and finish your earthly assignment that keeps calling for your attention right now.

There is so much more that I would like to have included in this book, including visualization, meditation, breath-work, neuroplasticity, heart/brain coherence, neuro-acoustics, neuroscience, Emotional Freedom Techniques, Owning Your Morning and How To Gain Mastery Over Life, Love, and Conditions and so much more.

But as the title suggests, this is "*The **LITTLE** Book of Quantum Secrets* –not the full version. So stay tuned for the full volume of this book. It is soon to follow.

Onward My Fellow Conquerors!!

APPENDIX A

7 SIMPLE STEPS TO SUCCESS IN THE QUANTUM FIELD

THERE ARE SEVERAL WAYS to engage the quantum field in such a way that you get what you want without delay.

Plan what you want.

Be very clear about what you want, why you want it, what it will mean when you achieve it.

Create a plan of action, even if you don't know all of the parts. The fact that you're writing it down and reading it is a great way to get really clear about what you want and

why you want it.

Speak into the universe with a clear, concise dream or goal that you hold and give it a timeline.

Walk in the direction of the goal, as if the fulfillment of your dream and goal has already happened. For example, if you want to build a business. And you've made up your

mind what you want to do. Then get an EIN number from IRS.gov and file the necessary paperwork with your state to create a corporation, LLC, partnership, nonprofit or sole proprietorship so that you can open a bank account.

Don't sit around thinking and praying and meditating only. Think about people who might get on your board of directors to help steer your business. Attend trade shows and other events that align with what you want. You have to baptize yourself in the space so that you can attract what you really need to succeed.

Don't wait for perfection. You have to get up and go! Now that you've done all of the preliminary work, ask for help. Seek guidance from God and man. Operate in the

vein of what you want and always be ready to reject any counterfeit representation of what you set in your heart to achieve.

Think of the quantum Field as your storehouse. And when you take the steps above, you will unlock its doors, gaining access to an upgraded life, abundance and wealth, a true love companion, and finally, the fulfillment of your purpose.

APPENDIX B

NAVAJO INDIANS AND THE BEAUTY PRAYER

Walking in Beauty

Closing Prayer from the Navajo Way Blessing Ceremony

IN BEAUTY I WALK.

With beauty before me I walk. With beauty behind me I walk. With beauty above me I walk. With beauty around me I walk. It has become beauty again.

Today I will walk out, today everything negative will leave me. I will be as I was before, I will have a cool breeze over my body.

I will have a light body, I will be happy forever, nothing will hinder

I walk with beauty before me. I walk with beauty behind me. I walk with beauty below me. I walk with beauty above me.

I walk with beauty around me. My words will be beautiful.

In beauty all day long may I walk. Through the returning seasons, may I walk.

On the trail marked with pollen may I walk. With dew about my feet, may I walk.

With beauty before me may I walk. With beauty behind me may I walk. With beauty below me may I walk. With beauty above me may I walk. With beauty all around me may I walk.

In old age wandering on a trail of beauty, lively, may I walk.

In old age wandering on a trail of beauty, living again, may I walk. My words will be beautiful...

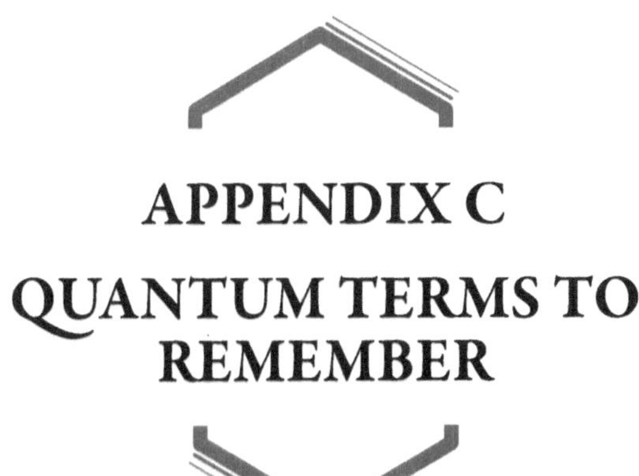

APPENDIX C

QUANTUM TERMS TO REMEMBER

QUANTUM MECHANICS EXPLORES the nature of reality at a fundamental level. It reveals that particles, such as electrons and photons, can exist in multiple states simultaneously—a concept known as superposition. As God is omnipresent, so do we have the ability to be in more than one place at the same time through the gifts of Spirit, including the word of knowledge, the word of wisdom, prophecy, and such like.

Additionally, particles can become entangled, meaning the state of one is directly connected to the state of another, regardless of distance.

In the quantum realm, uncertainty plays a crucial role, stating that we can't precisely predict both the position and momentum of a particle simultaneously. So stop saying things are impossible when they definitely are not. Things only become certain once you decide they are.

Quantum mechanics challenges classical intuitions and opens doors to mysterious phenomena like quantum tunneling, where particles can pass through barriers that had previously been considered impenetrable.

UNDERSTANDING THESE concepts can help you see the mysterious and interconnected nature of existence, giving us a deeper appreciation for the profound and awe-inspiring aspects of the universe and our role in it.

Finally, brethren, whatsoever things are true, whatsoever things are honest, whatsoever things are just, whatsoever things are pure, whatsoever things are lovely, whatsoever things are of good report; if there be any virtue, and if there be any praise, think on these things. **Philippians 4:8 (KJV)**

ABOUT

DR. PATRICIA VAN PELT-SCOTT is an eight-figure business owner, a serial entrepreneur and a personal growth strategist. In addition, she is an author, real estate developer, certified public accountant, and founder of The Season For Love Inc and Diamond VII Enterprises LLC, and co-founder and past-president of WaKanna For Life LLC.

Dr. Pat served four-terms as State Senator in the Illinois General Assembly. Before that, she was the founding CEO of a successful social movement organization where she served for 15 years.

Dr. Pat teaches The Advanced Life Series On Wealth, Health, and Love and Soulmate Attraction For Christian Singles. Patricia's goal in life is to advance civilization and to help solve world problems.

Currently, she is on a global tour with her *"Quantum Life: Get To The Top"* Conferences. Dr. Pat received her doctorate degree in Management of Non-profit Agencies after completing a dissertation on the study of Social Movements and Revolutions.

Dr. Pat is married to Gene Scott. They have four children. Dr. Van Pelt-Scott brings a wealth of experience and passion to her work.

DON'T MISS OUT!

Visit the website below and you can sign up to receive emails whenever PATRICIA VAN PELT-SCOTT, PhD publishes a new book. There's no charge and no obligation.

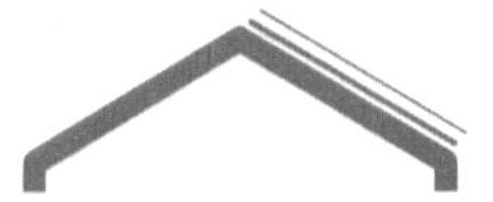

www.DrPat.net

bit.ly/littlequantumsecrets[1]

DID YOU LOVE *THE LITTLE BOOK OF QUANTUM SECRETS*: How To Get More Of What You Want On Purpose And Without Fail? Then you should read *God's Secret Storehouse* by P.T. Wilburn!

---

1. http://bit.ly/littlequantumsecrets

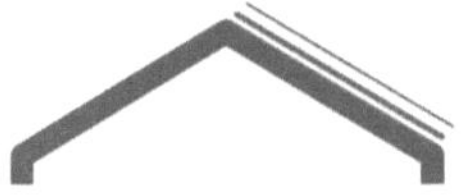

ARE YOU READY TO TRANSFORM YOUR LIFE AND STEP INTO A REALM OF BOUNDLESS ABUNDANCE AND SPIRITUAL FULFILLMENT?

IF SO, THEN THIS IS the perfect thing for you, unlock the secrets of unparalleled wealth and success in *"God's Secret Storehouse,"* where divine abundance meets life-changing prosperity.

Discover God's Secret Storehouse:

Experience Financial Freedom:

Learn practical strategies from the Bible and spiritual principles to attract wealth effortlessly and achieve financial freedom!

Inner Peace and Spiritual Fulfillment: Align your mindset and actions with God's promises for abundance, leading you to inner peace, joy, and spiritual growth!

Success Stories: Hear inspiring faith-based testimonials about individuals who have manifested extraordinary prosperity using the principles outlined in this transformative guide and how you can do the same!

Create Emotional Connection: Embark on a Journey of Transformation: Feel the excitement and empowerment of embarking on a transformative journey that touches your heart and soul. Imagine your dreams becoming a reality as you dive deep into the sacred wisdom of *"God's Secret Storehouse."*

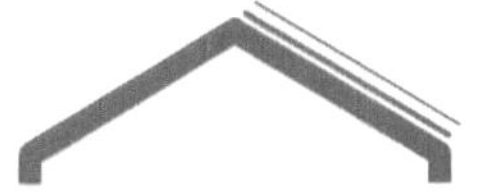

Now It's Your Turn to Get God's Blessing!

SEIZE YOUR ABUNDANCE Now: Don't wait for tomorrow; take action now to claim your abundance and manifest your dreams. Get Exclusive Bonus Content: When you get *"God's Secret Storehouse"* you will receive instant access to exclusive resources and bonuses that will accelerate your journey to prosperity.

Limited-Time Offer: Hurry, act now to unlock the heavenly vaults of divine abundance and create the life you've always dreamed of living.

The Windows Of Heaven Are Ready to shower unlimited wealth, fulfillment, and divine abundance, and it's all waiting for you!

Unlock Heaven when you get this book!

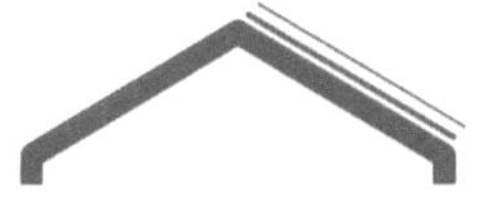

Also by PATRICIA VAN PELT-SCOTT, PhD

How to Attract the Man of Your Dreams: A Christian Woman's Guide to Success in Love

Don't miss out!

Visit the website below and you can sign up to receive emails whenever DR. PATRICIA VAN PELT-SCOTT, PhD publishes a new book. There's no charge and no obligation.

https://books2read.com/r/B-A-VLCMB-RWTID

BOOKS 2 READ

Connecting independent readers to independent writers.

Did you love *THE LITTLE BOOK OF QUANTUM SECRETS: How To Get More Of What You Want On Purpose And Without Fail*? Then you should read *God's Secret Storehouse*[1] by P.T. Wilburn!

ARE YOU READY TO TRANSFORM YOUR LIFE AND STEP INTO A REALM OF BOUNDLESS ABUNDANCE AND SPIRITUAL FULFILLMENT?

If so, then this is the perfect thing for you, unlock the secrets of unparalleled wealth and success in "God's Secret Storehouse," where divine abundance meets life-changing prosperity.

Discover God's Secret Storehouse:

---

1. https://books2read.com/u/b5JLX7

2. https://books2read.com/u/b5JLX7

Experience Financial Freedom: Learn practical strategies from the Bible and spiritual principles to attract wealth effortlessly and achieve financial freedom!**Inner Peace and Spiritual Fulfillment:** Align your mindset and actions with God's promises for abundance, leading you to inner peace, joy, and spiritual growth!**Success Stories:** Hear inspiring faith based testimonials about individuals who have manifested extraordinary prosperity using the principles outlined in this transformative guide and how you can do the same!

Create Emotional Connection:

Embark on a Journey of Transformation *Feel the excitement* and empowerment of embarking on a transformative journey that touches your heart and soul.*Imagine your dreams becoming a reality* as you dive deep into the sacred wisdom of "God's Secret Storehouse."

Now It's Your Turn to Get God's Blessing!

Seize Your Abundance Now: Don't wait for tomorrow; take action now to claim your abundance and manifest your dreams.*Get Exclusive Bonus Content:* When you get "God's Secret Storehouse" you will receive instant access to exclusive resources and bonuses that will accelerate your journey to prosperity.*Limited-Time Offer:* Hurry, act now to unlock the heavenly vaults of divine abundance and create the life you've always dreamed of living.

The Windows Of Heaven Are Ready to shower *unlimited wealth, fulfillment, and divine abundance,* and it's all are waiting for you!

Unlock Heaven *when you get this book!*

Also by DR. PATRICIA VAN PELT-SCOTT, PhD

THE LITTLE BOOK OF QUANTUM SECRETS: How To Get More Of What You Want On Purpose And Without Fail